The Moorish Essence of Pensacola

A Legacy of Exploration and Cultural Exchange from Estévanico to Tristan de Luna

Moorish March in Seville Press

The Moorish Essence of Pensacola

Jeremie Samuel

Published by:

Moorish March in Seville Press

P.O. Box 17315

Pensacola, Florida 32522

Moorish March in Seville is a historical and cultural-based community working to preserve the Moorish essence of Pensacola.

Visit our Facebook and Twitter: www.facebook.com/MoorsMarchInSeville, @MoorsMrchInSvll

ISBN: 978-1514370605

Contents

Introduction 1

The Moorish Essence of Pensacola 3

The Birth of the Commercial Empires of North Africa 18

  The Barca Family of Carthage 18

The Legacy of the Moors in Spain 22

Innovations of the Moors in Europe 27

The Moorish Origins of Navigation to the Americas 32

  The Moors Seek Refuge during the Crusades 36

Spanish Exploration: The Continuation of Inquisition and Conversion 39

Mustafa Zemmouri a.k.a Estévanico the Moorish Guide 43

Mustafa Zemmouri Arrives at the Seven Cities of Cibola 46

The Tristan de Luna Expedition 50

I Love Little Al-Andalus (Pensacola) 52

References 54

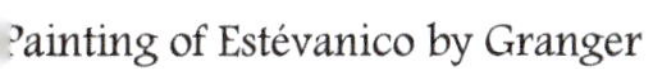

Painting of Estévanico by Granger

Tristan de Luna Statue at Plaza de Luna, Pensacola

Introduction

From 711 to 1492, the Moors carried the light of civilization to propel Europe out of the plagues of the Dark Ages into the European Renaissance. The Moors practiced high-end architecture; their temples surrounded by thousands of markets were at the heart of the cities and consisted of libraries and schools. Ninety-nine percent of the population in Christian Europe was illiterate while education in Moorish Spain was universal and available to the general public. The Moors established the world's first universities in North and West Africa. The University of Al-Karaouine is the first known university, and it is located in Fez, Morocco. This university originally was a mosque founded in 859 C.E. by a Moorish woman named Fatima al-Fihri. Located in the Northeast district of Timbuktu, Sankore University was in the Sankore Mosque. The scholarly chief judge of Timbuktu, Al-Qadi Aqib ibn Mahmud ibn Umar founded the Sankore Mosque in 989 C.E. During the 12th century, Sankore University had 25,000 students, in a city of 100,000 people.

Left: *The University of Al-Karaouine in Fez, Morocco* ***Right:*** *Sankore University in Timbuktu, Mali*

Accordingly, seventeen universities were established by the Moors in Spain and much culture was shared in the cosmopolitan provinces of Cordova, Seville, Almeria, Málaga, and Granada. The public library in the great city of Cordova kept 600,000 manuscripts. The city of Cordova was a magnificent city of the tenth century with well-paved streets that were illuminated by street lamps. The Moors cultivated the soils of Spain and grew gardens around their courts with flowers that held fine aromas. They developed hydraulic systems, irrigation with great water wells and built the first public baths in Europe. The Moorish culture was also one filled with romance and poetic literature. With the influx of these dark-skinned Moors into Spain and Portugal, the complexion of the Southern Europeans darkened while a festive and musical culture evolved. Those festive and romantic elements are all characteristics of the planet Venus. Whereas, Frenchman Robert Jastrow in 1686 compared the Moors to the people who inhabited the planet Venus stating:

"I can tell from here what the inhabitants of Venus are like; they resemble the Moors of Granada a small black people burned by the sun, full of wit and fire, always in love, writing verse, fond of music, arranging festivals, dances and tournaments every day."[i]

However, the Catholics proclaimed crusades against the Moors, persecuting them as infidels. Pope Urban II proclaimed the first crusade in 1095; other crusades followed in 1146, 1189, and 1211 which were all declared by Pope Innocent in Spain. The crusades strengthened the Christian empires and led to the union of Castile and León in 1230. Later, in 1469 Isabella reluctantly married her cousin Ferdinand of Aragon to bring about the union of Castile and Aragon. In 1479, they sanctioned the Inquisition, which denationalized many Moors by forced conversion into Christendom relabeling them as Moriscos.

Thousands of Moors were baptized into Christianity and servitude while Christians adopted many Moorish children as god-children and gave them Christian names. Some Moors converted to the religion of Christianity to remain in their homeland while others repatriated back to the kingdoms of Morocco, Mali, and Ghana.

Granada was the last stronghold in Moorish Spain. In 1483, the Moorish Prince Boabdil was taken captive and entered into a pact with Catholic monarch Ferdinand. In the aftermath, the Catholic Monarchs captured Málaga in 1487 and laid siege to the kingdom of Granada in 1490. Hence, in 1492 Boabdil surrendered Granada to the Catholic Monarchs, which was the fall of the Moorish rule in Spain known as the Reconquista.

While many Moors refused to live as Christian serfs and took cruisers back to North Africa, other Moors found refuge in the Americas. The evidence of Moors sailing to America before Columbus appears in the written and oral traditions of America. For example, the Nanticoke Moors of Delaware tradition tells that they are descendants of a crew of Moorish sailors who shipwrecked near Indian River Inlet, escaped to the shore, and intermarried with the aboriginal inhabitants. Furthermore, other Moors from Mali; the wealthiest empire of the middle ages traveled with the Moorish Prince Abubakari II to South America. In fact, the name Brazil comes from the Moorish tribe named Barasil.

The Moorish Essence of Pensacola

The environment and landscape of territories around the Gulf of Mexico are similar to the climate and atmosphere of Morocco and Spain; the countries that Moors inhabited. In fact, Pensacola lies on the 30th parallel north of the equator aligning with Morocco. The breeze that blows from the Gulf of Mexico to the shores of Pensacola is like the breeze that travels from the Atlantic Ocean to the shores of Morocco and Spain. Similar vegetation such as palm trees and citrus plants beautify Florida, Morocco, and Spain. Furthermore, the beaches on the coasts of Florida, Morocco, and Spain are gorgeous making these places home to Moors. Florida has a vibration and ecology similar to Morocco and Spain, which made Moors feel right at home. In essence, Florida was a Moorish land where Moors found a haven and brotherhood with aboriginals thereby bringing about unique cultures.

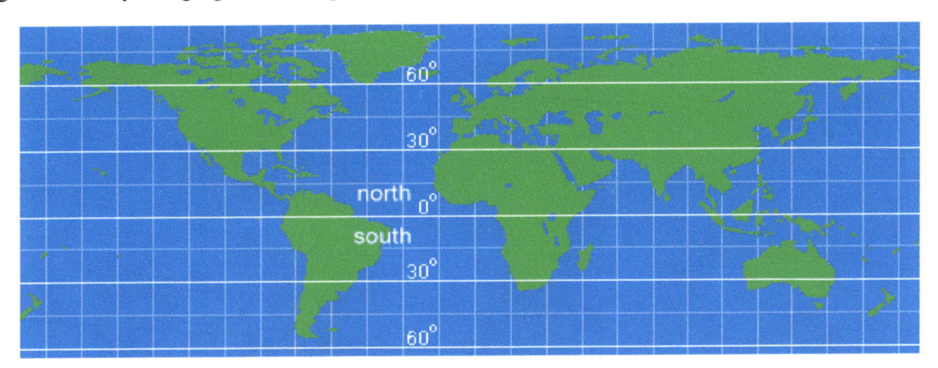

Pensacola aligns with Morocco on the 30th parallel north, 30 degrees above the equator

The mural of Estévanico shows him standing between Morocco and Florida. He came to Florida in 1527. (Mural by TatsCru in Azemmour, Morocco)

Agadir Bay and beach resorts of Morocco

Pensacola off the Gulf of Mexico where Trista de Luna landed in 1559 with 100 Moors

The Pensacola Museum of Art with Moorish style roofing used by the Moors of Northwest Africa and the Mediterranean region

Even upon the arrival of European colonists, the Wolf Trail or Pensacola Trading Path was still a lucrative trade route. Mustafa Zemmouri or Estévanico, the Moorish navigator and explorer, is celebrated today at the Estévanico International Festival in Pensacola. In Western perspective, he was the first person born in the continent of Africa to have arrived in the present-day continental United States. Mustafa Zemmouri was a guide and translator who lived in Azemmour, Morocco and traveled from Seville to Florida and Northwestern Mexico with Alvar Nuñez Cabeza de Vaca, Andrés Dorantes, and Alonso del Castillo Maldonado.

Tristan de Luna established the first Spanish settlement in North America and had 100 Moors on his fleet. The Moors founded the city of Seville in Moorish Spain, and it functioned as a clearing house to receive silks, gunpowder, leather and other goods from Africa and places around the Mediterranean Sea. Correspondingly, Seville is the oldest neighborhood in Pensacola.

Painting of Mustafa Zemmouri a.k.a Estévanico (Painting by Robert Rivera)

Three centuries later, Souanaffee Tustenukke or Chief Abraham served as a translator and diplomat for Chief Micanopy and the Seminole Nation. Tustenukke is a term given to a commanding officer in charge of defending the life and lands of the Seminoles. Chief Micanopy wears the Moorish-styled red turban with ostrich feathers in his portrait, and Abraham is also pictured wearing a turban. Many Seminoles wore fezzes also, as Chief Billy Bowlegs III is shown in a picture wearing a large maroon fez. Abraham lived in Pensacola but fled when Andrew Jackson arrived to colonize Florida and enslave freed and fleeing Moors also known as Maroons. Many people of African and aboriginal American descent fled from their homes in Pensacola with Abraham to "Negro Fort". Later in 1816 the fort was destroyed by U.S. ships. Abraham survived and went south around Tampa to find refuge at Billy Bowlegs' village.

Chief Micanopy, Chief Abraham and wife Hagan, Chief Billy Bowlegs III

However, before Moors like Mustafa Zemmouri and Souanaffee Tustenukke were in Pensacola, masses of Moors found refuge in the area during the Inquisition and Reconquista. Recorded as Spanish Africans and Creoles, these Moors settled in Pensacola to make a life for themselves based on Moorish heritage. The Moorish essence of Pensacola displays itself in the layout of the city. Pensacola's main districts are named Cordova, Seville, and Granada, which inspired me to term the city as Little Al-Andalus.

Like the Cordova of medieval Al-Andalus, the Cordova area is in the center of Pensacola serving as a commercial district with many businesses, restaurants, and commercial rental property. Cordova consists of Cordova Mall, Pensacola State College, and Sacred Heart Hospital. To the south of Cordova is Granada, a residential area populated by blacks–people of Moorish descent. Descendants of the first Moorish settlers established the Granada district. The community of Granada existed as a prosperous community up until integration that brought upon the disenfranchisement of its socioeconomic infrastructure like many other communities around the country at that time.

Cordova Mall in the Cordova Commercial District of Pensacola (Little Al-Andalus)

Cordova, Spain (Al-Andalus) functioned as the commercial and intellectual center of Al-Andalus. Cordova consisted of over 80,000 shops and a library of 600,000 volumes, schools, baths and the first streetlights in Europe. Its population consisted of Moorish, Spanish, and Arab including merchants and students from all over Europe, Africa, and Asia.

The Crown Plaza Hotel, originally the Union Depot located on the northeast corner of Wright and Tarragona Streets of downtown Pensacola, completed in 1882.

Left: *Painting of two Moorish merchants at the Alhambra by Rudolf Ernst (1854-1932)*
Right: *The Connoisseurs by Jean Discart (1856-1944)*

Cordova, Spain (Al-Andalus)

Cordova Commercial District of Pensacola (Little Al-Andalus)

The Alhambra, a Moorish Palace built in Granada in the 13th century, Al-Andalus

Historic Sacred Heart Hospital, opened in 1915, Granada, Pensacola (Little Al-Andalus)

South of Granada is the downtown historic district called Seville. Much like the Seville of Moorish Spain, it is a hub for art, music, and festivals like the Estévanico International Festival and the Festival of Five Flags. Similar to the Granada residential area, Seville was also historically a "black" district. In Seville; the oldest neighborhood of Pensacola, Julee Panton owned a cottage as a 'free woman of color'. Panton acquired the cottage for $300.00. She sold candles and pastries at the cottage. In June of 1809, Julee sold half of lot 65 to Angelica, a free black woman, for $41.50. Julee Panton sold the remaining half to Franciso Casini, a free Mulatto man for $212.00.[ii]

Julee Panton was in the population of 'free people of color' who lived in Pensacola under Spanish occupation. These 'free people of color' conducted business on Palafox Street until the Jim Crow law was implemented, and they were forced out. A large segment of the Pensacola downtown area consisted of Moorish descendants who built an economic infrastructure that produced prominent 'black' businesspeople. They established banks, building and loan associations, trust companies, pharmacies, insurance companies, schools, and grocery stores, etc. They resided in much of the waterfront property along Pensacola Bay like the Tan-Yards, Hawkshaw and Aragon Court.

Julee Panton Canton Cottage owned by a free woman of color in 1805, Historic Seville District of Pensacola.

Plaza de España, Seville

T.T. Wentworth Jr. Florida State Museum in Historic Seville District, Pensacola (Little Al-Andalus). Originally built as City Hall in 1907, the museum has Moorish style roofing like that the Moors brought into Spain from North Africa. The red roof is symbolic of the red caps (fezzes) that Moors wear with extraordinary pride.

Bars and Cafes after dark near Cathedral Square in Seville, Spain (Al-Andalus)

Bars and Restaurants at Seville Quarter in Pensacola (Little Al-Andalus)

For instance, D.J. Cunningham owned and ran the Excelsior Grocery Store, which was one of the best-known grocery stores and did the most business according to Booker T. Washington's book, *The Negro in Business*. John Sunday Jr., a Civil War veteran, also owned many properties in Pensacola including properties on Belmont and Devilliers built around 1875. Booker T. Washington recognized Sunday as a "wealthy man of color" for Sunday became wealthy through real estate and construction. Many houses Sunday built can still be seen as restored historical homes. His homes on Florida Blanca Street and West Romana are both restored historical structures. The sister of John Sunday Jr., Mercedes Sunday Ruby was the founder of St. Joseph's Catholic Church in 1877. John Sunday Jr. donated the land for the construction of the church during the Reconstruction Era. She was the daughter-in-law of Salvador Ruby, who was a Spanish-African (Moorish) commander of "free Mulatto" and "Negro" troops. They came to Pensacola in 1781 with the Governor of Louisiana, Bernardo de Galvez and helped the Spanish defeat the British. Ruby later went into business as a shoemaker in Pensacola, and also became the commander of an urban brigade of mulatto and black militia.

D.J. Cunningham and John Sunday. Photos from State Archives of Florida, Florida Memory

Moreover, another aspect of the Moorish essence of Pensacola is found in the lineage of Tristan de Luna, the conquistador who founded the first European settlement in North America at Pensacola. The settlement was called Santa Maria yet it was short lived due to its destruction by a hurricane. One hundred Moors came on De Luna's fleet to Pensacola and were left behind to fend for themselves after De Luna went back to Mexico. De Luna was born in the Moorish province of Borobia in northern Spain. He was of a Castilian family of the Borobia estates in the villas of Cicia and Borobia in Al-Andalus. The De Luna estates in Borobia yielded annually, the revenue of about three million maravedis.

Left: *Gold maravedi of the Almoravid dynasty in the 11th century* ***Right:*** *Almoravid gold maravedi coin from Seville, Spain, 1116*

Middle: *Moorish Castle in Borobia, Al-Andalus, Birthplace of Tristan de Luna* ***Bottom:*** *Flag of Tristan De Luna. Red Flag with white crescent moon resembles a flag of the Moors.*

Top: *Christ Episcopal Church of Pensacola built c. 1903 with Spanish Baroque architecture that evolved from the Moorish presence in Spain* ***Bottom:*** *U.S. Post Office and Federal Building with Moorish style roofing in downtown Pensacola (Little Al-Andalus).*

Maravedis are a former coin issued by the Moors in Spain. The word maravedi originates from the Arabic Marabitin; the Almoravids. The Almoravids were a Moorish dynasty of Morocco who formed an empire in the 11th century that stretched over the western Maghreb and Al-Andalus. They ruled in Spain from 1059 through 1147. Similar to Cristobal Colón and other European explorers, Tristan de Luna most likely received a Moorish education in his birthplace that allowed him to sail to America. In fact, the flag Tristan de Luna sailed under was red with a crescent moon in the likeness of a Moorish flag. Furthermore, Tristan de Luna's cousin Hernan Cortes was born in the Moorish kingdom of Granada in 1490. In 1537, Tristan de Luna was part of the Hernan Cortes expedition to find the Seven Cities of Cibola during the time Estévanico the Moor was on the path to the same destination.

***Left:** A possible depiction of Almoravid General Abubakari, riding a camel with a whip of knotted cords, from the 1413 chart of Mecia de Viladestes.* ***Right:** Portrait of Hernan Cortes, cousin of Tristan de Luna. He was born in Granada in 1490 and died in Seville on December 2, 1547.*

The Birth of the Commercial Empires of North Africa

When mentioning explorers arriving in Pensacola Bay and other regions of America, the contemporary historical perspective persuades most to think of the European explorers to the New World in 1492. On the contrary, the expeditions of European sailors were preceded by explorers of the Moorish empires. The Moors during their rule of Spain from 711-1492 produced innovations that led to the New World expeditions recorded in Western history. As inheritors of maritime science from the ancient Phoenician and Egyptian empires of North Africa; Moors were able to build commercial empires again through domination of the Mediterranean Sea and control of the Pillars of Hercules.

In Greek mythology, Hercules on his tenth labor crossed many countries to reach the western extremity of the Mediterranean. There he raised the two great columns, the mountains of Calpe and Abyla. These legendary Pillars of Hercules are today's Gibraltar and Ceuta. It came to be that whoever held jurisdiction over the Pillar of Hercules held power over the profitable maritime commercial industry of the Mediterranean and western passage to the Atlantic Ocean. The struggles between nations were to gain access to this water highway to increase commercial influence and activity; currency wars. So, it goes in the Greek story that Hercules engraved an entwined S-shaped symbol around the pillars that are Latin for 'Non plus ultra' ("Do not go beyond here"). This is the origin of a widely used monetary symbol that is used today as the dollar sign ($).

The Barca Family of Carthage

Carthage was the commercial power of the ancient world's trade industry in the Mediterranean. This powerful Phoenician city in northern Africa became the Mediterranean's most prosperous seaport with wealthy provinces; Sardinia, Corsica, and Sicily. The flags of Sardinia and Corsica with Moorish heads show the resonance of the early Moorish control over these lands and the Mediterranean. Holding sway over the Mediterranean and the seaports surrounding it produced a dominant position in the area of trade and commerce. However, the expansion of the state of Rome threatened the Mediterranean, commercial trade domination of Carthage. In the First Punic War, Carthage was stripped of its most valued province of Sicily. Rome also seized Corsica and Sardinia while the civil war took place in Carthage. In the third century B.C., Spain had lost the First Punic War to Rome 900 years after the Phoenician merchants are said to have founded Cadiz and Málaga.

During these events, a boy named Hannibal was born in 247 B.C. to the rich and influential Barca family of Carthage. He was the oldest son of Hamilcar Barca, the great Carthaginian general. Hannibal had two brothers and many sisters. His eldest sister married Hasdrubal the Great who had become his adopted brother. Hannibal's younger brother was named Mago. Hannibal's father Hamilcar Barca planned to capitalize on a staggering Spain and invaded the land after she lost in the First Punic War. In Spring 238 B.C. nine-year-old Hannibal went with his father as he left for war in Spain to gain revenge against the Romans and regain a foothold in Spain and Italy. It was then when the young Hannibal made the oath over an altar swearing that he would hate the Romans as long as he lived. It is said that Hannibal stood beside his father and stated, "I will swear that soon as age will permit I will use fire and steel to arrest the destiny of Rome."[iii]

Hamilcar invaded Spain and conquered a good part of southern Spain and the Valencian coastal area. To Hamilcar Barca, Barcelona owes its name and a second Carthaginian province was Carthago Nova or New Carthage, today's Cartagena. The young Hannibal Barca observed his father assemble a strong infantry of Numidian horsemen and Spanish soldiers with plans to conquer Italy. Through adolescence to his early manhood, Hannibal was exposed to the brutal scenes of warfare. He studied his father, the generals and all aspects of combat. Hannibal developed into a great Carthaginian warrior, an ancient annihilator that would eventually in his military career drive the Romans to defeat and despair. The death of Hannibal's father in 228 B.C. increased his hatred for the Romans. After Hamilcar's death, his son-in-law Hasdrubal was left in command of the army.

Hasdrubal was assassinated in 221 B.C., and some historical accounts say that he was killed in a gory gesture as the Romans threw his head in Hannibal's camp. Regardless, twenty-six-year-old Hannibal took over the army and in 218 B.C. seized and destroyed Saguntum. With a flaming Saguntum behind him, Hannibal looked ahead to implementing his father's plan to conquer Rome. Hannibal, the fierce commander of the Carthaginian army of 50,000 men and 37 elephants, launched his first attack against Rome to start the Second Punic War. The elephants provided better vehicles for combat than the former chariots of the Carthaginians. Like his father, Hannibal assembled a diverse army largely composed of mercenaries of different fighting styles. Hannibal led a cavalry through Spain over the treacherous Alps meeting marauding Gauls with boulders and arrow fire. They finally emerged with one-third of the army into northern Italy. Hannibal replaced the lost part of his army with Gauls and other recruited men on the way to Rome.

Hannibal's army consisted of a light infantry of sword-wielding men from Libya, a heavy infantry of charging Gauls and a heavy African and Spanish cavalry. Hannibal's coordinated attacks in diverse styles presented problems for the defense of the uniform Roman infantry. Hannibal and his army's first conflict was in the battle of Trebia against a Roman force led by Publius Cornelius Scipio, son of Scipio Africanus. In combat at the Ticinus River, 2,000 Roman soldiers were slain, and Publius Scipio severely wounded. In Trebia, Hannibal wore down the Roman infantry when he used the element of tactical surprise. The Carthaginian general ambushed the Roman force in a surprise attack that caused the death of 30,000 Roman soldiers at the bank of Trebia. Hannibal Barca drove his army south to meet two new Roman consuls Gnaeus Servilius and Gaius Flaminius. Assuming that no army would attempt to travel through the swampy, insect-infected Arno lowlands, the Roman generals split on both sides. Servilius and Flaminius blocked the eastern and western routes.

Unexpectedly, Hannibal took his army through the swampy Arno River though he would suffer losses including his eyesight from infection or an insect bite. Eventually, Hannibal again made it through a treacherous passage in the marshy lowlands of the Arno. Different from the previous rocky path of the Alps, the army faced no opposition in the Arno, but Hannibal still lost a large part of his force. After emerging from the lowlands of the Arno, the Carthaginian warrior Hannibal arrived in Etruria in the spring of 217 B.C. and set Tuscany; a Roman ally up in flames. Hannibal decided to send a message to the Roman allies and the Roman commander Gaus Flaminius. Hannibal's objective was to lure the hot-headed Flaminius into a pitched battle by devastating the region Flaminius had been sent to protect. After the attempt to lure Flaminius into battle by destroying the Roman allied region failed. Hannibal boldly marched around the Roman camp's left flank and decisively cut Flaminius off from Rome.

Hannibal's execution of this strategy was the first recorded turning movement in military history. This maneuver provoked Flaminius and Hannibal advanced through the uplands of Etruria and caught Flaminius in a narrow column of the shore of Lake Trasimenus. Hannibal destroyed the Roman army on the adjoining slopes, and Gaus Flaminius was killed. After the destruction of Tuscany and the defeat of two Roman infantries, Hannibal pushed forward to clash with a larger Roman army at Cannae. During the year 216 B.C. in Cannae, Hannibal orchestrated the most decisive defeat known in military history when he defeated perhaps the largest Roman army ever sent into battle. In the battle known in military history as the Battle of Annihilation, Hannibal and his army killed 70,000 Roman soldiers by sunset. In this battle, Hannibal executed another effective military strategy when he led his force towards the Roman opposition in a semi-circle formation. As the Roman army approached, Hannibal led his army into combat against their Roman enemy. After the two armies engaged in a struggle, the forefront center of Hannibal's semi-circle formation retreated. As the Roman force pursued them, the two sides of the semi-circle came forward to trap the Roman army in a V formation.

Shortly afterward, the Numidian horsemen rode behind the Roman army and sealed the triangular shaped trap. This military formation proved to be deadly, and Hannibal along with his army continued the massacre of the largest Roman army ever put into the field. Almost none of the men of Canae were spared, their women and children captured and sold as slaves. Among the large amounts of casualties were eighty members of the Roman city. The Romans reorganized themselves and coming to realized that they could not stop Hannibal in the head on head clashes they use a strategy of containment. The Romans would continually harass the Carthaginian army but never engage in a head-on battle.

In 209 B.C., Scipio Africanus captured Cartagena after engaging in battles against the Carthaginians in Spain. Therefore, in 205 B.C. the last Carthaginians were driven from the peninsula, and Hannibal was forced to return to Africa in 203 B.C. There in Africa, Hannibal was defeated by Scipio's army, and Scipio Africanus set up a Roman colony named after him by which the continent is known as Africa today. Although Hannibal never met his goals of destroying Rome, his military genius was such that his military tactics are still studied at military colleges. The defeat of Hannibal and the North African commercial power was the catalyst in the expansion of the Roman Empire. Over time, the Mediterranean became a melting pot of different people and cultures. The people of various cultures engaged in struggles for the economic advantages of the Mediterranean regions.

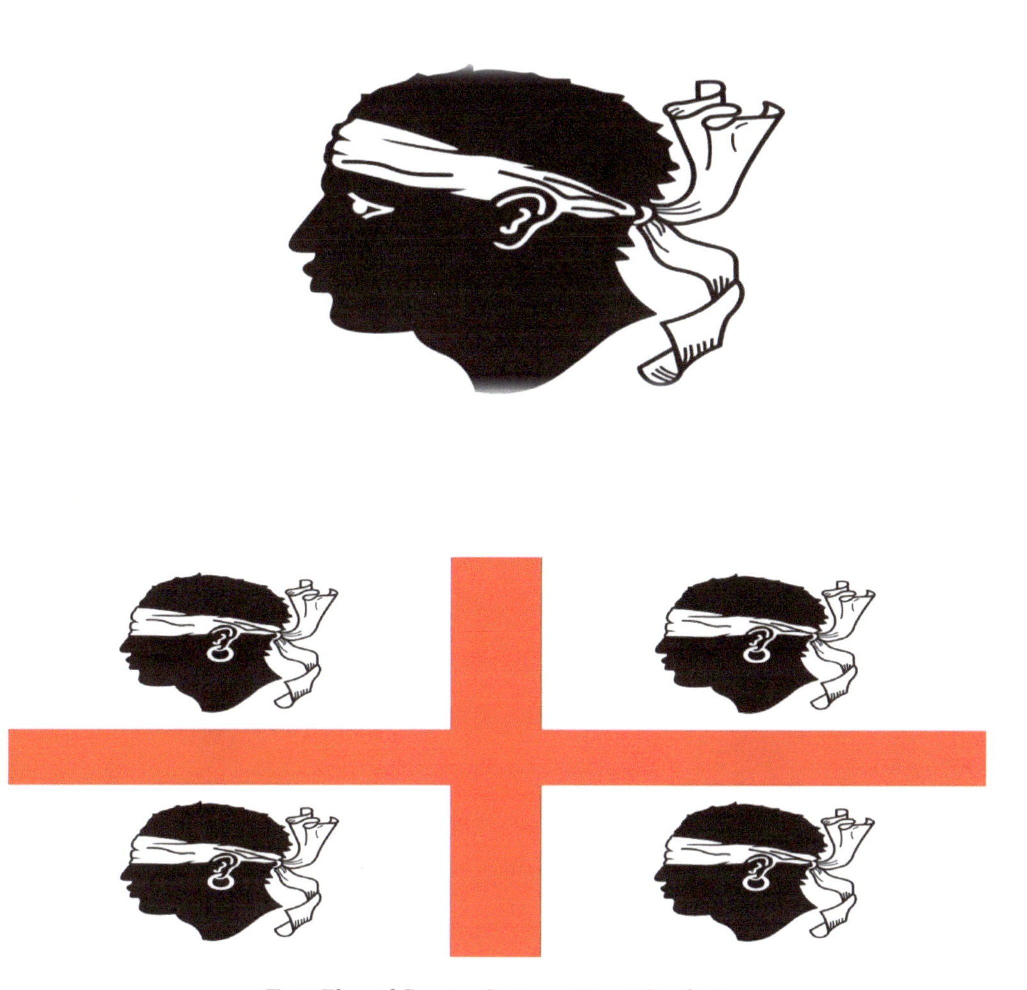

Top: *Flag of Corsica Bottom: Flag of Sardinia*

The Legacy of the Moors in Spain

It was in the year 710 C.E., a century after the Prophet had begun his message and the promulgation of Islam, the Moors envisioned building a vibrant civilization in the fertile land of Spain. The Moors began to plan for the re-establishment of their influence in Spain and domination of the Mediterranean. The Moorish front united under the banner of the emerging religion. By this time, they were in a position that gave them the potential of defeating the Byzantine navy and again gaining control of the eastern and southern Mediterranean basin. Due to the sociopolitical crisis that was taking place in Spain, the rivals of the Byzantine King Roderick were surging against him. Musa Ibn Nusayr was inclined to undertake the conquest of Spain.

Thus, in 710 Musa Ibn Nasyr assigned the officer Tarif to an exploratory mission to the Iberian Peninsula. Tarif headed four hundred men and a group of 100 cavalries in the successful occupation of the southern tip of the Peninsula. In the southern tip of the Peninsula, the city of Tarifa bears his name. He left Tarifa and returned to Africa with an impressive amount of booty. Consequently, tariff–today's term for rates of duties or taxes on imported goods is derived from the Moorish officer Tarif. When Musa received the news of Tarif's successful exploratory mission and impressive acquisitions, he was ready to further the pursuits of conquest.

After a good year of preparation, in 711 the Moabite governor of Mauritania, Tarik ibn Ziyad was commissioned to lead a Moorish army into Spain. Tarik, the Moorish general, left Tangier leading an army of ten thousand which consisted of nine thousand Moabite soldiers. The Moorish soldiers crossed the straights and landed on the Iberian Peninsula at a fortress that from that very moment became known as Jabal Tarik or Gibraltar (Tarik's Mountain, Rock of Gibraltar). Gibraltar became the base of Moorish operation. The Visigoth king of Spain was King Roderick. Count Julian sought revenge upon Roderick of Ceuta, who had allegedly raped his daughter Florinda. Julian led an army and backed a side in the many struggles between Gothic factions that occurred before the invasions of the Moors. In addition to the struggles within the country, the population was affected by plagues and the Goths had lost the warrior instincts of the barbaric hordes that invaded Rome. In the tenth century, the Spanish General Chronicle states that the Goths were all thin and sickly "for they had just gone through two awful years of pestilence and famine".[iv]

Though the Germanic barbarians were successful in the destabilization of Rome, they lacked some of the foundational aspects of civilization such as agriculture, education, and sanitation. Thus in the Dark Ages, plagues hit the whole of Europe. Spain like most of Europe had no public libraries, universities, lights, running water or baths. While the European nobility possessed the skills of reading and writing, the masses were illiterate. All this would change when the Moors came to power. General Tarik, head of the Moabite army pushed to the north from Gibraltar. When informed of the African invaders; King Roderick quickly assembled an army of forty thousand to a hundred thousand men to march against the Moorish force. Tarik was a determined leader of his soldiers who were inspired by the religion of Muhammad. The Moorish force was able to unite the brave soldiers of different Moabite tribes of Africa under the banner of Islam.

At the outset, it was victory or death for Tarik. Like Hannibal, the great North African general of antiquity who sought to defeat the Roman army at any cost, Tarik placed the same price before his zealous African infantry. It is said that Tarik after landing burnt the ships they came on to show the army that there was no turning back. Tarik would rally the Moorish warriors before they marched into battle against the European army. Looking into the eyes of the veiled dark faces of the Moabite warriors, Tarik would speak words of inspiration to galvanize them on the way towards conquest. In the summer of 711, he is said to have roused his troops stating:

"My brethren, the enemy is before you the sea is behind; whither would ye fly? Follow your general; I am resolved either to lose my life or to trample the prostrate king of the Romans."[v]

Tarik and the Moabite army defeated a Gothic army as they marched from Cartagena to Cordova. They continued mounted upon their swift horses to meet the Visigoth army in the plains of Jerez on the banks of the Guadelete. Roderick was reportedly "wearing on his head a golden crown, encumbered with a heavy robe of silken embroidery, and reclining on a litter or car of ivory drawn by two white mules, as was the way the Gothic kings of those days went about."[vi] This hindering dress of King Roderick especially in military activities reflects the perspective of historical commentary on how the Goths had lost the warrior spirit of their predecessors in luxury and complacency of peaceful circumstances.

"The Goths were no longer the victorious barbarians, who had humbled the pride of Rome, despoiled the queen of nations, and penetrated from the Danube to the Atlantic Ocean. Secluded from the world by the Pyrenean mountains, the successors of Alaric had slumbered in a long peace: the walls of the cities were moldered into dust: the youth had abandoned the exercise of arms; and the presumption of their ancient renown would expose them in a field of battle to the first assault of the invaders."[vii]

In the accounts of the 13th-century Castilian ruler Alfonso X, he praises the Christian Visigoth kingdom as a powerful, religious, and honorable kingdom. Alfonso X goes on to lament in how the Visigoth kingdom fell when King Roderick opened the charmed palace and violated the sacred palace thereby serving as an opening for Moorish invasion. Alfonso X explained the misfortunes of the Visigoth kingdom downfall as being inflicted by the devil.

"The devil, who is the enemy of humanity and in whose envy there always remained the search for doing it harm, sowed the evil and black seed in the kingdom of Spain, and caused pride in the powerful; laziness and negligence in the religious men; discord among those who had enjoyed peace and love; luxury and great lust among the rich and the well-to-do; complete indifference among the wise and people of understanding; to such an extent that he made the bishops and the clerics like vile men of the people, and kings and princes like thieves. It was in this manner, as we have said that the kingdom of the Goths of Spain was destroyed, a kingdom that was great and large, so great that its dominion lasted long and stretched from sea to sea, to the city of Tangier in Africa to the Rhone River."[viii]

At the end of that passage, Alfonso X mentions that the Gothic kingdom stretched to the city of Tangier in Africa. Tarik governed this region, and it was the launching pad for the Moabite army in the conquest of Spain. Alfonso X continues to honor the history of the Visigoths and again lament that such a powerful, rich, and noble kingdom could be overthrown and destroyed by internal conflict at the hands of the "Moros".

"All the Moorish soldiers were dressed with silk and black wool that had been forcibly acquired; the reins of their horses were like fire; their black faces were like pitch and the most handsome of them was like a cooking pan; thus, their eyes shone like flames, their horses fast as leopards and their knights were more cruel and more harmful than a wolf in the midst of a herd of lambs at night. The vile people of Africa, who were not used to kindness and all of whose deeds were accomplished with tricks and deceits and who were not used to protect but to exploit great riches, are now exalted... "Poor Spain. Your death was so afflicted that none remained to lament you in your dire suffering, for you are [considered] more dead than alive."[ix]

In his accounts, Alfonso X shows the aggressive and speedy pursuit of the Moors in the conquest of Spain in 711. In the battle of Guadalete, the Moors engaged in combat with a much larger Visigoth army. The Moorish army engaged in a series of violent hit-and-run attacks against the lines of Visigoth soldiers. The two armies engaged in conflict for hours. They were in a long period of a standstill until the Visigoth army weakened in part by an army led by Count Julian. A cavalry wing of the Visigoth army that secretly pledged with Julian to rebel against Roderick stood aside giving the Moabite cavalry an opportunity to exploit an opening left in the Visigoth force. This event would be the turning point in the battle and lead to a victory for the Moorish army. The battle was a bloodbath that lasted seven days.

The Visigoth army suffered an extremely high amount of casualties. King Roderick's death is uncertain; he is said to have drowned or escaped. He may have been slain in the final hours of battle. It is also said that he leaped from his car mounted on his swift horse Orelia and fled from the field of battle. He escaped a soldier's death only to drown ignobly in the waters of the blood filled river. After the Moorish army had routed the Visigoth army, Tarik wasted no time in continuing the conquest of Spain. The Moors led by Tarik marched north swiftly conquering several cities on the way. Using no time in celebration, Tarik immediately sent contingents against Málaga, Elvira, Murcia, and Cordova. The Moorish cavalry, facing light opposition reached Toledo through Jaén. They found a deserted city in Toledo, except for Hebrews who would later help the Moors build a citadel of learning in that city.

There in Toledo, Tarik attained a splendid bounty of gold, silver, precious stones, vestments, crowns, horses and many other prizes. Among these prizes was the Table of Solomon which was reportedly made of solid gold and silver and decorated with precious stones. Certainly, these prizes urged Tarik to find more rewards in conquest. Running on the adrenaline of conquest, in relentless vigor Tarik and the tireless Moabite army advanced into Castile, Leon, and the Asturias. The Moors had spread themselves too thinly throughout the Peninsula. Therefore, after overthrowing the European dominance of Spain within a month's time, Tarik sought no further conquest. Content with the conquered territories, the Moors settled and started to work the lands they had acquired from the southern tip of Spain to the northern city of Toledo.

The city of Toledo would provide a strategic position for the Moors against the Christian powers of the north. Toledo was the first base of Moorish operation in Medieval Spain. In Toledo, Tarik supervised the Hebrews and delegated the administration to the natives of the conquered territories. The conquest of Spain had been officially executed successfully by the Moabite General Tarik, and his Moabite army fortified itself in Islam. The Moabite division of the Moorish power has been categorized as Hamitic Berbers. The Hamitic branch deriving from the biblical story of the sons of Ham: Kush, Mizraim, Put, and Canaan. Musa Ibn Nasyr may have been jealous of the Tarik's remarkably swift achievement in the occupation of Spain. Ironically, this sentiment was felt by the Arabs for the Moabites who were the most integral arm in the Moorish dominion of Spain. In fact, it was the mass participation of these Moabites in the conquest of the Iberian Peninsula that caused the Europeans to term the migration of Islamic people to Spain as 'Moros'.

The terms for the two divisions of Moorish Spain were Moabite (African) and Hagarite (Arab). Consequently, it was the Moabite side that planted the first seeds to flourish into an empire of marvelous culture consisting of prosperity, learning, and grand architecture. The Moorish empire extended from Al-Andalus southward to the Niger River (Senegal) and out beyond the Atlantic Ocean in the Americas. Moorish dominions rooted in advanced culture, intellect and commercial industry sprouted up in Africa from Morocco, Mauritania, Mali, and Ghana. The Moabite dynasties successfully spread Islam among the people of Ghana, Mali, Mauritania, and elsewhere. This expansion would play an important role in the waves of Moabites known as Almoravids and Almohads, who came out of Morocco and Mauritania into Spain during the 11th century. The Almoravids and Almohads came into Al-Andalus to solidify the high point of Moorish dynastic rule there in the eleventh and twelfth centuries.

Alcazar Palace of Seville, Spain, built by the Almohad Moorish Dynasty that ruled in Al-Andalus 1121-1212 A.D.

Innovations of the Moors in Europe

The Moabite preponderance in Spain changed the appearance of the population in the land. Under the domination of the Moors, Al-Andalus represented Paradise as a result of the Moorish ability to cultivate the land. The Moabites could transform even the most meager lands into gardens of delight, with perfumed gardens and constant flowing water. When Arabs and Syrians came after the conquest was complete; many Moabites went to cultivate the chilly heights in the north. It was the Moorish mastership of irrigation that would be of most benefit to them in Al-Andalus. The import of orange and palm trees from Africa and other plants into Spain by the Moors produced a potent mixture of Southern Europe and Africa in the makeup of Al-Andalus.

In 785, Abd ar-Rahman I brought the Christian Cathedral in Cordova to build the Great Mosque of Cordova in its place. The Great Mosque of Cordova became the heart of Moorish Spain and was a cultural landmark and commercial center. A market composed of eighty thousand shops circled around the elaborate Cordova mosque at the apex of its prosperity. This temple complex was the blueprint for most Moorish cities where cultural, intellectual and commercial activities ensued around the perimeter of the temple. Moreover, Cordova, Seville, Granada, Málaga, Jaén, and Almeria functioned as economic centers for merchants and goods were imported from all regions of the Mediterranean world and beyond. These cities provided markets where Moorish and foreign traders could engage in international trade. Cities that were off the coast of the Mediterranean Sea were in prime positions for the profitable maritime trade of the middle ages.

Cities such as Almeria, Seville, and Málaga functioned as commercial clearinghouses where Muslim, Jews, and Christians handled imports.[x] In medieval, international trade of the Mediterranean region, merchants were grouped by religious affiliations and in some instances they only traded with other merchants of the same religious affiliations. However, during the medieval period of Moorish rule on the Iberian Peninsula, Muslim merchants formed the dominant trade group. It was common to sight Moors with turbans and fezzes cruising in the Mediterranean Sea transporting goods into the peninsula while other Moors were sailing in the Atlantic with goods and animals from West Africa.

In Al-Andalus, scholars converged in the cities like Seville, Granada, Málaga, Toledo, Denia, Almeria, and Badajoz to enjoy patronage from the Moorish kings. Mathematics, astronomy, law, philosophy, and metaphysics were some of the subjects studied at the Moorish intellectual centers. From the 8th to the 15th century, the Moors carried the light of civilization in the Peninsula and essentially laid the foundations that brought about the European Renaissance. In Al- Andalus, Moors introduced algebra, algorithms, chemistry, paper, and botany to Europe.

The old word for arithmetic, 'algorism', is derived from the Moorish mathematician Musa al-Khwarizmi like the word 'logarithm'. Adelard of Bath (c. 1090 – c. 1150) introduced translated Arithmetic of al-Khwarizmi and introduced Euclid to the Western World. Later Robert of Chester (c. 110 – c.1160) transcribed another work of al-Khwarizmi, the Algebra. Furthermore, the word 'algebra' derives from the name of the polymath, Al-Jabir el-Hayyan (c. 721 – c. 815). Al-Jabir, as an alchemist, engineer, geographer, philosopher, physicist, pharmacist, physician, astronomer and astrologer represented the multi-facet aspects of Moorish scholarship.

The civilization of the Moors was universal. The fields of alchemy, astronomy, marine science and geography were greatly influencing the West during the Moorish rule of Spain and Portugal. One instance of remaining traces of Moorish influence is that a quarter of the words in the Spanish language are from Arabic roots, the language by which the Moors unified. Some of the words imported into the English language from Arabic are algebra, almanac, apricot, cipher, sherry, sugar, zenith, and zero. Some of the chemical terms transferred are alkali, alcohol, antimony, borax, elixir, syrup, talc, theodolite, and tarter. Aldebaran, Altair, Betelgeuse, Rigel, and Vega are some of the stars with Arabic names.

Text like the *Liber Mahameleth, the Ikhtisar al-jabr wa' l-muqabala*, written by Abu Abd Allah Muhammad b. 'Umar b. Muhammad b. Badr before 1343 and treatises on the division of inheritance ('ilm al-fara ' id) provides examples of mathematics in Moorish Spain. *The Liber Mahamelith* deals with arithmetic and algebra, ending with a long collection of practical problems that may be of interest to a merchant. Archimedes, Euclid, Nichomachos of Geras, Abu Kamil Shuja 'b. Aslam al-Mizri (the Egyptian) and Muhammad b. Musa al-Khawarizmi were quoted in the book. *The Ikhtisar* is a treatise on elementary algebra dealing and indeterminate equations in the Diophantine tradition. Likewise, Moorish mathematician, Abu ' l-Hasan 'Ali b. Muhammad al-Basti al-Qalasadi wrote extensively on arithmetic and algebra. He was influenced by the works of the Moroccan mathematician, Abu ' l-Abbas Ahmad b. Muhammad, known as Ibn al-Banna' al-Marrakushi (1256-1321).

The designs of Moorish architecture are largely built on geometric patterns and in Moorish Spain there were also works on geometry and spherical geometry. The discovery of an important written work called Al-Istikmal by al-Mu' taman of Saragossa contained four incomplete manuscripts containing fragments of the work dealing with number theory, plain geometry, study of the concepts of ratio and proportion following books V and VI of Euclid's Elements, the geometry of the sphere and of other solid bodies and conic sections. According to Julio Samso, the extant parts of *the Istikmal* proves that al-Mu' taman had an important royal library containing the best books available in the 11th century for the study of higher mathematics: *Euclid's Elements and Data, Archimedes' on the Sphere and Cylinder*, the books on Spherics by Theodosius and Menelaos, Appolonius' Conics, Ptolemy's Almagest, Thabit b. Qurra's treaties on amicable numbers and Menelaos' theorem, the treatise of the Banu Musa on the measurement of plane and spherical figures, Ibrahim b. Sinan's book on the quadrature of the parabola and Ibn al-Haythams Optics, etc.[xi]

Indeed, great emphasis was placed on the pursuit of knowledge in Al-Andalus. The numerous mosques served as educational centers for anyone who wanted to be educated. Umayyad emir, 'Abd al-Rahman II was very influential in the establishment of a literary tradition that continued at the Umayyad Court in succeeding generations. The Umayyad Court and first capital of Al-Andalus was Cordova; the nerve center of politics and the major intellectual center of Al-Andalus. Many students traveled to Cordova from all over the country to study at the feet of the great teachers. Moorish rulers who appreciated knowledge surrounded themselves with men of letters and encouraged intellectual activities. 'Abd al-Rahman II secured books from the East and had an appreciation for religious and secular sciences, music, and poetry.

The brilliant musician Ziryab came to Cordova by the patronage of 'Abd al-Rahman II. Ziryab was a Moor known as Pájaro Negro (Blackbird) in Spanish. He was a singer, poet, composer, and musician. According to historians, Ziryab was well known for his black color and singing voice resulting in the nickname 'Blackbird'. Ziryab was also a polymath having knowledge in culinary arts, fashion, geography, astronomy, meteorology, and botany. His career flourished in the 9th century at the Moorish dominion of Cordova where he brought about innovations in fashion, music and culture. Ziryab is considered the founder of Andalusian music for his unique style of songwriting and music performance which influence Spanish music for generations. He used an eagle's beak instead of a wooden pick to the guitar played by the Moors in Spain named the oud. Ziryab added a fifth pair of strings to the oud and dyed the four colors representing the four humors and the fifth representing the soul. Zyryab is a quintessential example of the Moorish Renaissance Man that developed innovations in Europe bringing the region out of the Dark Ages.

Yahya al-Ghazal (d. 865) who was known for his sharp tongue and quick answers and 'Abbas Ibn Firnas, a man of unusual abilities were some other scholars in the court of 'Abd al-Rahman II. Ibn Firnas constructed a simulated sky with lightning and thunder and also invented a formula for manufacturing crystals.[xii] Cordova reached its apex as an intellectual center in 929 – 961 under Caliphs 'Abd al-Rahman III and his son al-Hakam II. Leading intellectuals of the day: Ibn 'Abd Rubbihi, al-Qali, al-Zubaydi, and Ibn al-Qutiyah among others enjoyed patronage from 'Abd al-Rahman III and al-Hakam II. The Cordovan calendar and the expansion of Cordova mosque started under the reign of al-Hakam II. He also made education available to anyone who desired it by founding many schools. Scholars came to cities like Seville, Granada, Toledo, Denia, Almeria, Badajoz, Valencia, and Málaga to receive endorsements from rulers.

Astronomy and astrology were sciences closely related and financed by the party kings who set up to twelve independent states of Al-Andalus. The *qadi* (judge) of Toledo, Ibn Sa' id, (1070) published writings titled *Kitab tabaqat al-umam (Universal History of Sciences)* which gives insight on the development of science at this time. A contemporary of Sa' id was the alchemist Abu Maslama of Madrid who wrote a book entitled Rutbat al-hakim, (the Rank of the Wise) which contains descriptions of his inventions, formulae and instructions for purification of precious metals. In the course of his experiments, he was also one of the earliest alchemists to record the usage of mercurix oxide and first to note the principle of conservation of mass. As an alchemist, mathematician, astronomer, and economist Abu Maslama was among the most brilliant of Moors in Al-Andalus during the party king period.

The Moors also worked in the field of astronomy and astrology from the period of the caliphate to the rule of Muluk al-Tawa'if in the 11th century and the reign of the Moabite dynasties during the 12th and 13th centuries. Julio Samso states that 'Abd al-Wahid b. Ishaq al-Dabbi (c. 800) is probably the first Andalusi astrologer to have left a written work. Al-Dabbi composed an astrological urjuza of which only 39 verses are extant and in which astrological predictions are based on the Late Latin "system of the crosses" (tariqat ahkam al-salub). The Moors assimilated traditional Latin and Hellenistic astrology. The astrological system makes predictions based on the positions of the seven planets in the four triplicities of air, water, fire and earth and also account for the astrological houses with the signs of the zodiac.

The Moorish institutions placed much emphasis on the development of astronomical instruments like the sundial, astrolabes, equatorium and analog computers. The astrolabe is an ancient astronomical instrument used to show how the sky looks at a specific place at any time. Universal astrolabe seems to be of Moorish origin. The Andalusi polymath Abu 'l-Salt Umayya b. Abi 'l-Salt (c. 1067 – 1134) wrote a treatise on the astrolabe in Alexandria (Egypt) in 1109-10. The astrolabe solved the problem of the division of the houses necessary to cast a horoscope and other problems related to time and the position of the Sun and stars in the sky. As astrology became a profession, the equatoria came about and solved the graphical problem that the tedious process of computing planetary longitudes using a set of astronomical tables in astrolabes.

Ibn Al-Samh's equatoria used a set of plates, one for each planet with another for the epicycles which were kept within the mother of an astrolabe. Furthermore, the longitude of a planet was measured using a scale engraved on the rim of the instrument as an ecliptic scale.[xiii] The Moors continued to improve these astronomical instruments which help advance their progress in other sciences like agriculture, navigation, and even architecture. J. Vernet wrote that medicine, like botany, is linked to astronomy owing to the interest of pharmacologist in obtaining "simples", that is to say, plants which might be employed as remedies, without adulteration.

Top Left: *Silver-inlaid brass planispheric astrolabe Spain, probably Toledo, 14th century Engraved copper alloy inlaid with silver.* ***Bottom Right:*** *Silver-inlaid brass planispheric astrolabe from Moorish Spain, 14th century. Astronomers, navigators, and astrologers use astrolabes.*

The Moorish Origins of Navigation to the Americas

The astronomic tools of the Moors were the foundation of the early Moorish expedition back to the West and European expedition to the Americas thereafter. Abdul-Hassan Ali Ibn Al-Hussain (871-957 C.E.) wrote in his book *Muruj adh-dhahab wa maadin aljawhar (The meadows of gold and quarries of jewels)* that during the rule of the Muslim caliph Abdullah Ibn Mohammad (888-912 C.E.) a Moorish navigator from Cordova named Khashhash Ibn Saeed Ibn Aswad sailed from Delba (Palos) in 889 C.E., crossed the Atlantic to reach an unknown territory (ard majhoola) and returned with treasures. The Americas were referred to as the unknown territory in a large area in the ocean of darkness and fog in Masudi's map of the world. Furthermore, another Moorish navigator named Ibn Farrukh sailed from Kadesh into the Atlantic in February 999 C.E. during the reign of Hisham (976-1009 CE). Abu Bakr Ibn Umar Al-Gutiyya, a Moorish historian, narrated that Ibn Farrukh landed in Gando (Great Canary Islands) visiting King Guanariga and continued westward where he saw and named two islands, Capraria and Pluitana. He arrived back in Spain in May 999 C.E. [xiv]

The famous Muslim geographer and cartographer Al-Sharif Al-Idrisi (1099-1166 C.E.) wrote in his book *Nuzhat al-mushtaq fi ikhtiraq al-afaq (Excursion of the longing one in crossing horizons)* that a group of seafarers from North Africa sailed into the sea of darkness and fog (Atlantic Ocean) from Lisbon of Portugal in order to discover what was in it and what extent were its limits. These Moors finally reached an island that had people and cultivation and on the fourth day, a translator spoke to them in the Arabic language. There are illustrations of Portuguese Moors in the new world diplomatically interfacing with indigenous Moors in Brazil. Therefore, it was the Moors who were the first Portuguese explorers and started the practices of intermarrying with the natives of the Americas and its islands, a practiced that was emulated by later Portuguese explorers.[xv]

The journey of Shaikh Zayn Eddin Ali Ben Fadhel Al-Mazandarani across the sea of fog and darkness is well-documented in the Muslim reference books. His journey started from Tarfaya, South Morocco during the reign of King Abu-Yacoub Sidi Youssef (1286-1307 C.E.) 6th ruler of the Marinid dynasty. Al-Mazandarani voyage lead him to Green Island in the Caribbean Sea in 1291 C.E. Another Moorish historian Chihab Ad-dine Abu-l-Abbas Amad Ben Fadhl Al-Umari (1300-1384 C.E.) described in detail the geographical explorations of Mali's sultans in his famous book *Massaalik al-abasaar fi mamaalik al-amsaar (The pathways of sights in the provinces of kingdoms.)* These documents recorded the journeys of different Moorish navigators and explorers in expeditions to the Moorish provinces of the Americas. [xvi]

Likewise, the word Moor can be traced back to the Phoenician word *mauharin* for Western. The land farthest west is America or Al-Marrakanus where the early people of ancient America were the Amaru. The Americas, Northwest, and Southwest Africa were within the ancient transoceanic empire in which the Moors controlled passageways out into the Atlantic Ocean. In West Africa, the land is called Mauritania; the town of the Maurs. The original name for Ancient Egypt was Tamauray which denotes the land of the Maurs. Therefore, the Greeks called the African tribes Mauray which transferred into the Late Greek *mauroi*, 'black'.

From ancient times to medieval periods the Moors were synonymous with maritime science which resulted in the word *mer* being rooted in the words dealing with maritime and commerce. For example, merchants, marine, commerce, and admiral are all words that originate from the Moors and their activities upon bodies of waters. The word admiral for the captain of maritime expedition comes from the word *Amir*, Arabic for leader and *al* meaning *the*. Therefore, admiral was translated into English from Moorish navigators who were referred to as Al-Amir or Amir-al, leader of the fleet; people. These admirals studied at intellectual centers in Moorish dominions like Cordova, Seville, Granada, and Jaén.

Likewise, one of the largest intellectual and commercial centers of the Moors was in the Moorish kingdom of Mali where Mandinka-speaking Moors studied various fields of science and mathematics. The dominions of the Mandinka-speaking Moors like Jenne and Timbuktu were highly functional learning and commercial centers. Many merchants and scholars worked in Jenne and Timbuktu. Jenne was a commercial center and intellectual center surrounded by a network of waterways. For 800 years, Jenne attracted merchants and scholars of the Maghrib. Likewise, Timbuktu owed its importance to its geographical position. It was a meeting place for travelers by both land and sea. Nomadic people of the desert, riverian people of the Niger River, and European explorers and merchants stopped in Timbuktu.

During the medieval period of Moorish rule of the Iberian Peninsula, the Moors formed the dominant trade group. They were known to transport tigers, leopards, exotic birds and even elephants and giraffes to distant lands. The horses the Moors traveled on became known as Barbary horses and were highly valued. The Moors of West Africa thrived within industry and commercial trade largely based on the gold trade. Precious commodities like gold and diamonds sustained a thriving trade that had been established long before the middle ages. The Mali Empire, Songhay Empire, and Guinea established the gold trade before the coming of any foreigners. The rich economy in West Africa financed voyages of Moorish explorers across the Atlantic where they brought gifts, settled and married into native tribes in the Americas.

The Mali Empire of Soundiata, Abubakari II and Mansa Musa in the 14th century and the Songhay Empire of Ghana under Sunni Ali Ber and Askia Muhammad in the 14th and early 15th century sprouted out of the prosperous trade of West Africa. E.W. Bovill states in his book *The Golden Trade of the Moors* that the Moors were able to form the permanent growth of Muslim scholarship and wealth which brought all benefits of propinquity to seats of learning and cultural ease. In the middle of the eleventh century, the Western Sahara was in a state of political excitement over the triumphant progress of the Moors. The Moors of West Africa enjoyed access to the Atlantic Ocean and established trade routes with other empires on the golden coast along with trade relationships with the Maghrib and Al-Andalus. For example, Askia Muhammad allied himself with the scholars of Timbuktu to usher in a golden age of Muslim scholarship in Songhay. Mansa Musa of Mali also built relationships with his Moorish contemporaries in Maghreb and Andalus.[xvii]

The West African nobility and merchants were in touch with the larger world of Islam which improved diplomatic causes and international trade relationships. Like Al-Andalus, Islam in West Africa was adopted and synthesized with the traditional customs of their countries to bring about their unique version of Islamic culture reconnected to the African root from which it came. Essentially, the religious affiliations of the Moorish empires of West Africa positioned them amongst the medieval, multi-national trading network of affluent Islamic commercial powers.

For example, Mansa Musa assigned the Andalusian poet and architect named as-Sahili to build a mosque for worship in his homeland. The foundation of the mosque was built of burnt bricks, the use of which is still unknown. Furthermore, Mansa Musa set out on the Hadj in 1324 accompanied by a host of followers numbering 500. Musa was mounted on a horse ahead of a splendid caravan with gifts of gold in 80 to 100 camel loads each weighing 300 pounds. Mansa Musa showered gold everywhere he went on his way to Mecca. His 500 followers each carried a staff of gold weighing 8 ounces. Despite the Malian emperor's extraordinary display of wealth, it was his piety and open-handed generosity that made him popular in Cairo and other places. As a result of this spectacular pilgrimage, the name of Mansa Musa became known throughout a large part of the civilized world. In the first quarter of the 14th century, the fame of Mali spread through Europe and the Middle East and across the Atlantic Ocean.

The New World was not new to the ancient Moors, and many traveled the Atlantic and Pacific Ocean hundreds of years before the first European ships touch the shores of the Americas. For example, the name of Brazil is derived from Moorish voyagers who named an island they discovered near that region after the name of their tribe, the Barasil. It was a name comprised of the two noble titles of Bey and El (Beyrasel). This expedition was set forth by the predecessor of Mansa Musa, the Moorish Sultan of Mali named Abubakari II, who sent forth a fleet of 200 ships across the Atlantic. These expeditions were easily financed due to Mali being the richest empire in medieval times.

In fact, Mansa Musa who inherited the crown of Mali from Abubakari II in 1311 is recorded as the richest human ever at a net worth of 400 billion. During the time of Abubakari II and Mansa Musa, Timbuktu was a major trading center that allowed the empire to accumulate much wealth. Therefore, Abubakari II with the assistance of Moorish ship builders from Egypt and Mali constructed vessels off the coast of Senegambia in Africa. Abubakari II had 2,000 ships, 1,000 with the finest men, sorcerers, physicians and navigators and the other 1,000 ships were loaded with food supplies to last the crew for two years.

Medieval author al-Umari wrote the following account of the Abubakari's expedition by Mansa Musa:

The ruler who preceded me did not believe it was impossible to reach the extremity of the ocean that encircles the earth (here meaning the Atlantic); he wanted to reach that [end], and was determined to pursue his plan. So he equipped two hundred boats full of men, and others with water, gold and provisions, sufficient for years. He ordered the captain not to return until he had reached the other end of the ocean, or until he had exhausted the provisions and water. So they set out on their journey. They were absent for a long period, and at last just one boat returned. When questioned, the captain replied: "O Prince, we navigated for a long period, until we saw the midst of the ocean a great river which flowed massively. My boat was the last one; others were ahead of me, and they were drowned in the great whirlpool and never came out again. I sailed back to escape this current." But the Sultan would not believe him. He ordered two thousand boats to be equipped for him and for his men, and one thousand more for water and provisions. Then he conferred to regency on men for the term of his absence and departed with his men, never to return or show any sign of life. In this manner I became the sole ruler of the empire.[xviii]

It is believed that the great river which these voyagers witnessed was the Amazon, but there is also evidence of Malian Moors in pre-colonial North America on the Mississippi River. Who did these Moorish settlers meet? They were the aboriginals of America who had similar complexions and facial features and were the descendants of an international culture of ancient and advanced civilizations, making it natural for the Moors from the east to marry into their nations.

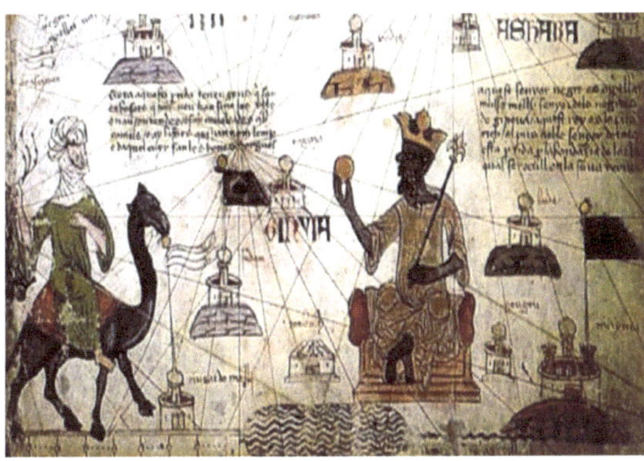

Left: *Artists rendition of Abubakari II* ***Right:*** *Picture of Mansa Musa and Moorish dominions of Mali*

In 1312, Mandinka Moors of Mali arrived in the Gulf of Mexico and explored the American interior via the Mississippi River. Anthropological evidence has proven that under Mansa Musa's instructions, the Mandinka explored parts of North America by the Mississippi and other river systems. According to Barry Fell, there was an iron mask of West African origin that was discovered in Londonderry, Ohio by Fred Wayble and initially discounted as "a piece of metal that bears a resemblance to a human face."[xix] Writings at Four Corners, Arizona depict the Mandinka bringing elephants from Africa.

The history of Moorish voyagers traveling to the 'new world' and marrying into the indigenous tribes of America is preserved in the oral history of tribes like the Nanticoke of Delaware who were of the Lenape and thereby part of the Iroquois Confederacy. Tradition says that the Nanticoke are descendants of a crew of Moorish sailors who shipwrecked near Indian River Inlet and escaped to shore and intermarried with the indigenous people living there. Evidence of the Moors amongst the indigenous people of America exists in the Mandinka numbers of the Nanticoke numerical system. Frank Speck sites Dr. Briton in the case of influence of Mandinka Moors on the Nanticoke language. He writes:

"African influence, however, is shown most conclusively by the fact that a list of Nanticoke numerals (1-10) recorded by Pyrlaeus, a missionary to the Mohawk in 1780, proved, through the investigations of Dr. Brinton, to be Mandingo or a closely related West African language."[xx]

Another part of the Nanticoke tradition says that the Moors were pirates from the Spanish Main. The Moors are attributed to starting piracy after the expulsion in 1492 to the last expulsion in 1609 when they took captives from Christian dominions for ransom in retaliation to the Inquisition. Maritime traffic increased in these times while the Moors were in naval wars with European nations in the Mediterranean Sea and America off the coast of the Atlantic. The wars transpired up until the 1830's in the United States of America and became known as the Barbary Wars. Frank Speck was inclined that the major Moorish admixture with the natives off the coast of Delaware occurred during these times.

"I am inclined to credit the general claim that Moorish sailors might have been shipwrecked on the treacherous shoals off the southern Delaware coast and come ashore to the shelter of the Indian natives. When this might have happened it is difficult to say, unless we assume that it was during the years of piracy on the high seas in the seventeenth and the early part of the eighteenth century (1650-1720 approximately). Importance of the term "Moors" in connection with the pirates of the West Indies suggest relationship in this case."[xxi]

The Moors Seek Refuge during the Crusades

However, the Moors of the Maghreb and Al-Andalus through control of access to the Atlantic were able to escape to America when the conflict came between Moorish powers and Christian powers of Europe. The Moors had settled in Spain founding royal cities and building societies virtually unfettered by Catholic monarchs for nearly four centuries until Pope Urban proclaimed the first crusade in 1095. Thereafter, the Almoravids rose in North Africa and established reinforcements for the Moorish front in southern Europe as they captured Santarem, Badajoz, Porto, Evora and Lisbon of Portugal in 1111.

The Moors established fraternal orders that housed various knowledge and disciplines serving as schools of thought and enlightenment. This foundation laid by the Moorish orders led to the enlightenment period which would lead to the European Renaissance. The Moorish orders inspired the Christian fraternal orders, firstly the Knights Templars, founded in 1121 and its coats of arms bore three Moors' head. As the internal conflict grew in the Moorish empire between the Almoravids and the Almohads, the Christian kingdoms started to unite. Officials of these Christian kingdoms translated the knowledge that they inherited from Moorish archives through the crusades and converted Moors known as Moriscos.

Five years after the founding of the Knights Templars, Archbishop Raymond founded the first school of translators in Toledo from 1126-1151. In result, the Koran was translated into Latin in 1143 and the second crusade took place from 1146-1148. The third crusade took place in 1189 which allowed the Catholic kingdoms to gain more leverage and information thereby the charters of the Universities of Oxford and Paris came in 1200 and the Magna Carta in 1215. Christian kingdoms were uniting as Castile and Leon did in 1230 and the Anglo-Portuguese Alliance of 1373. Meanwhile conflict escalated in Morocco and Al-Andalus leaving many Moors in a dilemma of staying in Spain to deal with conversion during the Inquisition and possible expulsion.

On the other hand, Ghana, Mali, and Senegal received an influx of Moors who were being pushed out from the northern dominions. Some of the Moors went into Asia. However, the next option was the land farthest west or western Maghreb; Al-Marrakanos–America where Moors had traveled and established provinces since the ancient times. Delacy O'Leary, a British Orientalist, spoke about the area of "Western Maghreb" extending beyond the Atlantic during the pre-Columbian Islamic era. Ferdinand of Aragon and Isabella of Castile married in 1469 which lead to the union of Castile and Aragon in 1479. In the following year, the Catholic Monarchs established the Inquisition in Spain. During the last century of Moorish rule, the Moors of Al-Andalus and Africa sailed to the Western Hemisphere settling in the Americas. Barry Fell in *Saga America* presented evidence of an earlier Moorish presence in America through similar architecture and shared language within North American tribes. As the Mandinka influence exists in the numerals of the Nanticoke, the Iroquois shared a language with North African-Iberian language. Furthermore, just as there were Islamic schools in Northwest Africa and Spain, traces of Islamic schools are present in Medieval America.

While Moors were repatriating into Africa and settling in America, the Moorish dominions of Spain were in decline. In 1482, Boabdil had taken the throne from his father Mulay Hasan in Granada. As the Catholic Monarchs focused on Málaga and the last Moorish stronghold in Granada, Boabdil was taken captive at Lucena where he entered into a pact with King Ferdinand. Thereafter, the Catholic Monarchs captured Málaga in 1487 and laid siege to the city of Granada in 1490. The fall of the last Moorish kingdom of Spain came in 1492 when Boabdil surrendered Granada to Isabella and Ferdinand. With jurisdiction over the former Moorish provinces, the Catholic rulers were able to catalog the information from the archives of Moorish knowledge and study the vast wisdom of the Moors.

Pictures of Boabdil, Sultan of Granada, the last Moorish Kingdom of Spain

The Court of Lions in Granada's Alhambra Palace

Spanish Exploration: The Continuation of Inquisition and Conversion

The Catholic sovereigns ordered Prior del Prado, a later Archbishop of Granada to form a council of geographers who studied Cristobal Colón's proposal to sail to the Indies. Cristobal Colón was the original name of the explorer who became known as Christopher Columbus. During the time of Columbus' exploration proposal, the Catholic monarchs were preoccupied in engagements with the Moors in the war of Granada. The royal commission deliberated his proposals for five years, and Isabella promised that she would hear his proposals after her mind was free from preoccupations with the war against the Moors.

Determined thereafter, Columbus went to the Court of the French King to make his proposal. However, he felt strong ties with Spain and considered himself a Spaniard having spent much time there and was convinced by Fray Juan Perez that his enterprise meant much to Spain. Santangel, a Jewish converso and financier gave powerful support to Columbus and offered to provide funds to support his enterprise. Isabella was persuaded by these last intersessions, and Christopher Columbus was sanctioned by Isabella to travel to the Indies in 1492. He learned from the Moors about the foreign lands and what they possessed while studying under them in Seville and Granada. This fact is cited in the accounts of Columbus' son Ferdinand who mentioned a letter written in 1501 to the Catholic Sovereigns:

"Very High Kings,

From a very young age I began to follow the sea and have continued to do so to this day. This art of navigation incites those who pursue it to inquire into the secrets of this world. I have passed more than forty years in the business and have traveled to every place where there is navigation up to the present time. I have had dealings and conversations with learned men, priests, and laymen, Latins and Greeks, Jews and Moors, and many others of other sects...During this time I have made it my business to read all that has been written on geography, history, philosophy, and other sciences."

Lisbon, Portugal and Guinea of West Africa are two of the places where Columbus traveled to study the navigational sciences. The Moors had ruled Portugal from the 8th to the 15th century as they did in Spain. When the Inquisition and expulsion of the Moors and Jews took place in 1492 and 1496, Moors took sail from Portugal and landed in America where they referred to themselves in many cases as Portuguese. Before Christopher Columbus went to Spain to propose his enterprise he spent much time in Portugal. It is in Lisbon where Columbus made his home and married Dona Felipa Monez whose father was Pedro Moniz Perestrello, a notable seafarer. He learned this while living amongst his father-in-law's widow where she told him that the King of Portugal had licensed him and two other captains to discover new lands. Sailing southwest they discovered the islands, Madeira and Porto Santo.

Ferdinand wrote in his accounts that Columbus received Perestrello's sea charts.

"Seeing that her stories of these voyages gave the Admiral much pleasure, she gave him the writings and sea-charts left by her husband. These things excited the Admiral still more; and he informed himself of the other voyages and navigation that the Portuguese were then making to Mina and down the coast of Guinea, and greatly enjoyed speaking with the men who sailed in those regions. To tell the truth, I do not know if it was during this marriage that the Admiral went to Mina or Guinea, but it seems reasonable that he did so."[xxii]

Therefore, being in Spain, Portugal, and Guinea of West Africa, Columbus new of the Moorish navigation and was aware of the Moorish provinces in the Americas. It was from Granada after Boabdil had surrendered the last kingdom of the Moors where Columbus received the capitulations from Ferdinand and Isabella. He departed on May 12, 1492, from Granada to Palos. Columbus along with early Spanish and Portuguese navigators benefited from Moorish geographical and navigational information on voyages across the Atlantic. In fact, Columbus had two captains of Moorish origin during his first voyage across the Atlantic. Martin Alonso Pinzon was the captain of the Pinta, and his brother Vicente Yanez Pinzon was the caption of the Nina.

The Pinzon family was related to the Abuzayan Muhammad III who was the Moroccan Sultan of the Marinid dynasty from 1362-1366 C.E. It was on August 1492, that the fleet of three small ships, the Nina, the Pinta, and the Santa Maria set out on their voyage from the port of Palos. On October 12, 1492, Columbus reached a little island in the Bahamas that was called Guanahani by the natives which he renamed San Salvador. Guanahani derived from Mandinka and modified Arabic words, Guana meaning 'brother'. Moreover, Ferdinand wrote about the Moors of dark-skinned natives his father saw in Honduras.

"The people who live farther east of Pointe Cavinas, as far as Cape Gracious a Dios, are almost black in color."[xxiii]

Moorish natives known as Almamy also lived in the same region at the time. Almamy in Mandinka and Arabic languages was the designation of Al-Imam or Al-Imamu, the chief of the community or a member of the Islamic Imami community. Columbus coasted along from island to island and landed on the coasts of Cuba and Haiti, which he named Espanola where natives told him that Moors had landed before his arrival. They presented the spears of these Moors to Columbus. The spears were tipped with a yellow metal that the natives called Guanin which derived from the West African word meaning 'gold alloy'. Guanin is also related to the Arabic word 'Ghinaa' which means 'wealth'. After having the metal tested back in Spain, Columbus learned that the metal was 18 parts gold (56.25%), 6 parts silver (18.75%) and 8 parts copper (25%), the same ratio as the metal produced in the African metal shops of Guinea.[xxiv]

Leo Weiner in his book, *Africa and the Discovery of America* wrote that Columbus was well aware of the Mandinka presence in the New World and that the West African Muslims had spread throughout the Caribbean, Central, South and North American territories, including

Canada, where they were trading and intermarry with the Iroquois and Algonquin Indians.[xxv] The shared culture manifested in the dress and veils worn by the Iroquois women in addition to women of the Southwest and Caribbean. Their veils bear semblance to those worn by women in Granada and Morocco at the time of Columbus.

Columbus had knowledge of this headdress due to his time in Granada where Moorish women wore similar headdresses and handkerchiefs. On his third voyage in 1498, Columbus' crew went ashore and found natives of South America using colorful handkerchiefs of symmetrically woven cotton. The Admiral noticed that the handkerchiefs resembled the headdresses and loincloths of Guinea in their style, colors, and function. He referred to them as Almayzars, which is an Arabic word for wrapper, cover, or apron which was the cloth Moors imported from West Africa into Morocco, Spain, and Portugal. Ferdinand called the native cotton garments 'breechcloths' of the same design and cloth as the shawls worn by the Moorish women of Granada.

Granada means 'people of the hills' and at the beginning of Moorish rule in Spain some of the Moors from Morocco, Ghana, Mali, and Senegal settled in the mountainous areas and built housing there. These Moors made up the majority of the army of 10,000 that conquered Spain with General Tarik in 710. They were known as the Moabites. Ivan Van Sertima points out that Ges, Zamoras, Marabitine, and Marabios are names that have transcontinental influence. Marabitine and Marabios relate to Moabite or Marabout; the Holy Men and Women of the Moorish Empire. The Marabouts were the protectors of the Moorish provinces and often functioned as buffers against Catholic-Albion encroachment. Moorish castles and mosques were built upon mountains in Spain, and similar architecture existed in the Americas during pre-Columbian periods and thereafter. Christopher Columbus saw people of the Carib as “Mohammedans” and admitted in his papers that he saw a mosque on the top of a beautiful mountain while sailing near Gibara on the northeast coast of Cuba.

In the beginning and end of his voyage, Columbus had fortunate and less fortunate encounters with the Bobadilla family. The Bobadilla family name is derived from the Arabic Islamic name Abou Abdilla, and the family was related to the Abbadid dynasty of Seville (1031-1091 C.E.). In the Canary Islands which were called Gomera, Columbus fell in love with Beatriz Bobadilla, daughter of the first captain general of the island. However, Columbus was put in chains and transferred from Santo Domingo back to Spain in November of 1500 by another Bobadilla; Francisco, the royal commissioner. Contrary to contemporary beliefs, Columbus never made it to the continent of America but ended his navigational career as a prisoner.

Oil Painting of Nobles of Peru, Don Francisco de Arabe, Sons Pedro and Domingo by Andres Sanchez Galque, 1599. Artwork from 'Revealing the African Presence in Renaissance Europe'. National Museum of Prado, Madrid, and the Walters Art Museum.

Left: *Painting of a Moroccan Girl by William Merritt Chase (1849–1916).* ***Right:*** *The Moorish Chief (originally titled The Guardian of the Seraglio) by Eduard Charlemont, 1878. Philadelphia Museum of Art*

Mustafa Zemmouri a.k.a Estévanico the Moorish Guide

Mustafa Zemmouri, the Moor from Morocco found friendship with the indigenous people of the western states when he arrived in the region with Cabeza de Vaca. He travelled to Pensacola on the way to the Seven Cities of Cibola. Mustafa Zemmouri corresponded well especially with some the Zuni and Pueblo tribes due to close relations to the architecture, culture, dress, and cosmology with his home of Morocco and the Dogon of Mali. When Cabeza de Vaca, Alonso del Castillo, Dorantes and Mustafa Zemmouri arrived at the Pueblo villages, Cabeza de Vaca noted the large earthen homes, complexion and build of the people along with the social practices of these natives. Cabeza de Vaca states that they were 'the most docile people we met in the country, of the best complexion, and on the whole well built.'[xxvi]

Castillo and Mustafa Zemmouri went with women as guides who took them to the Rio Grande. They wore cotton robes that were of a finer quality than those at New Spain. They had coral beads from the South Sea and turquoises from the north. They presented Dorantes with five emerald arrow points that they used in their ceremonies. The natives told Cabeza de Vaca that they traded for the emerald arrow tips for feather-bushes and parrot-plumes with people from some very high mountains from the north where there were villages with many people and mansions. The cliff-side Pueblo villages of the Southwest resemble the earthen homes of the Dogon tribe in Mali that they built in cliffs.

On the subject of their social practices, Cabeza de Vaca noticed, "we found the women better treated than in any other part of the Indies as far as we have seen. They wear skirts of Cotton that reach as far as the knee and over them half-sleeves of scraped deer-skin, with strips that hang down to the ground, and which they clean with certain roots, that clean very well and thus keep them tidy. The shirts are open in front tied with strings; they wear shoes."[xxvii] Like the Moors who came in the centuries before to share the knowledge and set up schools, Mustafa was able to integrate into the society of the natives. He learned much amongst them becoming an interpreter between the natives and the Spaniards. The related elements between Zuni and Moorish culture of Morocco and Mali created a situation for Mustafa Zemmouri to assimilate into the indigenous cultures of the Southwest just as his Moorish predecessors did before. Cabeza de Vaca noted the special communication between Mustafa Zemmouri and the Southwest aboriginals,

"It was the negro who talked to them all the time; he inquired about the road we should follow, the villages–in short, about everything we wished to know."[xxviii]

It was Mustafa Zemmouri's communication with the natives that allowed him to bring back food for the starving Spaniards. Fields of maize, beans, and gourds stretched behind the hills where the natives lived in the earthen houses. Mustafa Zemmouri learned secrets of the land that he did not share with the Spaniards. He became the center of early American discovery legend after the Spaniards left and he stayed behind to explore the Seven Cities of Cibola.

At the beginning of June 1536, Cabeza de Vaca and the wanderers reached Compostela in Mexico and were brought before Governor Nuno de Guzman. Guzman greeted the visitors graciously giving them fine clothing which the men were unable to wear after going naked or clad in skins on their journey. The governor listened to their accounts of their journey with deep interest for they searched the lands beyond the mountains that he had wished to discover.

Guzman still believed the land to be rich in gold and silver, and Cabeza de Vaca's accounts confirmed the Legend of the Seven Cities. After two months in Mexico City, Cabeza de Vaca decided he would return to Spain and set sail in the spring of 1537. Castillo married a wealthy widow and retired peacefully into a private life. Dorantes was offered a small expedition by the Viceroy, which Don Antonia de Mendoza accepted and then withdrew thus the enterprise was never executed. Mustafa Zemmouri and natives of New Mexico were the only people left with the Viceroy. The Viceroy reported the situation in a complaint to the Emperor: "I have spent a great amount of money on the expedition, but for reasons I do not know it has all come to nought. From all the preparation I have made there are left to me only a Negro who came with Dorantes, some slaves whom I have purchased, and some Indians, natives of the country, whom I have assembled."[xxix]

The natives from New Mexico and Mustafa Zemmouri would play vital roles in the quest for the Seven Cities of Cibola. The Viceroy assigned a Franciscan friar named Marcos de Nizza to the voyage. Zemmouri and the Indian guides would assist Friar Marcos in exploring the lands to the north of Mexico. Don Hernan Cortes was the rival of Guzman and also had the zeal for the conquest of the Seven Cities. His ships were the first to cross the Pacific and reach the Spice Islands. Cortes dispatched two ships lead by Diego Hurtado de Mendoza to explore the Pacific coast. Hurtado de Mendoza went up as the coast 27 degrees north before he turned back. Fortuna Ximenes was put in command of the second expedition sanctioned by Cortes. Fortuna Ximenes landed in a small bay where he and twenty of his crewmen were slain by the Pacific Coast natives. Afterwards, Cortes embarked on his very own expedition to find the island of California which the legend concerning the place circulated amongst explorers. A third of Cortes crew were Moors including his second-in-command, Juan Garrido. Garcia Ordonez de Montalva published *The Adventures of Esplandian* almost two decades after the Columbus voyage. It describes California and its inhabitants stating:

"On the right hand of the Indies is an island called California, very close to the Terrestrial Paradise, and it was peopled by Black Women without any man amongst them, for they lived in the fashion of Amazonia. They were of strong and hardy bodies, of ardent courage, and of great strength. Their island was the mightiest in the world, with its steep cliffs and rocky shores. Their arms were all of gold, and so was the harness of the wild beasts which they tamed and rode."[xxx]

The name California derived from the warrior queen named Califia, who ruled over the kingdom of Black Women living on the mythical Island of California. Califia came from the Arabic word *khalif* meaning ruler or leader which became known as *califa* in Spanish. The word was transferred to Spain when the Moors ruled Southern Europe. Furthermore, the Spanish were accustomed to being ruled by dark-skinned Moors and influenced by the literature and culture they brought with them. In the legend, Califia fought on the side of Moorish warrior leaders to help them regain Constantinople from the Christian armies who held it.

According to John William Templeton, Califia is an example of a genre of literature from the 1300s to the 1500s featuring black women as powerful, wealthy and beautiful. The legend describes Califia as the most beautiful of a long line of queens who ruled over the mythical realm of California. She is a regal black woman that is brave, tall and strong limb, full in the bloom of womanhood. Califia led a fleet of ships that demanded tribute from surrounding lands. Califia is depicted as the Spirit of California and is displayed in modern-day murals, sculpture, films and paintings.

She is symbolic of the untamed and bountiful land prior to European settlement. As Califia plays a role in the mythic origins of California, legends often have a connection to historical reality. For instance, most of the early depictions of California's first inhabitants show dark-complexioned aboriginals. Many of the paintings from the San Franciscan Mission show beautiful black natives with strong limbs as mentioned in the legends. In May 1535, Cortes landed in what he believed to be the legendary island and named the place California.

Left: *Bust of Queen Califia* ***Right:*** *Ohlone tribe of California Ceremonial Dance by Expedition artist Jose Cardero 1816*

Mustafa Zemmouri Arrives at the Seven Cities of Cibola

After Cabeza de Vaca's misfortunes in Southeastern territories of North America and the Island of Ill Fate along the Texas coast, Hernando De Soto would take the opportunity to colonize Florida. De Soto was born in Vila Nueva de Barcarrota at the very end of the 15th century or beginning of the 16th century. He had gained much respect in the conquering of Peru, receiving honor amongst his contemporaries as one of the bravest captains to land in the West Indies. De Soto was one of the wealthiest men to return to Spain. At his first appearance at court, he spent a thousand dollars in his share of Inca spoils.

Hernando de Soto received capitulation from the emperor on April 20, 1537, giving him authority to "conquer, pacify, and people" the territory of Florida.[xxxi] The capitulation also gave De Soto governorship of Cuba. De Soto was to embark within a year's time. He would independently finance his enterprise of conquest, he gathered 500 men for the expedition. Cabeza de Vaca landed at Lisbon and gave his account of his journey across North America. He told his kinsfolk that the region he explored was "the richest Countrie in the world". Cabeza de Vaca advised them to sell their goods and go with De Soto. Hernando de Soto called for Cabeza de Vaca to join him on his expedition for the knowledge he gained on his voyage would be of great benefit to his enterprise. De Soto presented an enticing offer to Cabeza de Vaca, who initially accepted the offer and then reversed his agreement with memories of the disastrous outcomes of the Narváez voyage still fresh on his mind.

On April 6, 1538, Hernando de Soto embarked on the voyage from San Lucar, Andalusia. His fleet consisted of seven large vessels and three small vessels. The fleet reached Cuba in the later part of May. De Soto collected more materials for the expedition including horses raised by the Cubans. In Cuba, De Soto recruited more men for his army and visited all the post on the island. The army increased to six hundred men who were all armed and brought to the port of Havana from where the fleet of nine vessels would embark. This enterprise was more of conquest than exploration. They set sail for Florida on Sunday, May 18, 1539.[xxxii]

Meanwhile, Frier Marco de Nica had already departed from the town of Saint Michael in the province of Caliacan on Friday, March 7, 1539. He was accompanied by Frier Honoratus, Andre Dorantez, Mustafa Zemmouri and natives of the Southwest regions to function as guides in the search for the Seven Cities of Cibola. Frier Marcos de Nizza sent Mustafa Zemmouri to go north to seek any news of rich country that was of great importance. Frier De Nizza asked that if Zemmouri found rich country that he not go further but return in person or send Indians with a sign. It was settled upon that Zemmouri would send a white cross one hand's length for an average matter, a cross two hands' length for a great matter and for a country greater than Mexico he would send a great cross.[xxxiii]

Zemmouri left Frier De Nizza to search for the cities of Cibola in spring of 1539. Mustafa Zemmouri as a guide in the search for the Seven Cities of Cibola entered the Zuni village of Hawikuh in New Mexico. In just four days after his departure, Mustafa Zemmouri sent messengers with a great cross as tall as a man back to Frier Marcos de Nizza. Mustafa found brotherhood amongst the Zuni tribe assimilating within their culture. Roberts and Roberts wrote, "…still others suggest that Estavan, who was black and wore feathers and rattles, may have looked like a wizard of the Zuni."[xxxiv] The Zuni villages resembled the villages of Morocco creating a smooth transition for Mustafa Zemmouri to assimilate.

He became a priestly figure amongst the indigenous people he met in the Southwest. Zemmouri traveled flamboyantly, dressed in bells and feathers attached to his arms and legs. A handful of beautiful women accompanied Zemmouri on his journey. Other natives who believed they could travel with impunity under the protection of Zemmouri carried his provisions. Zemmouri carried with him turquoise and other gifts he received from the natives throughout his journeys. He also carried a gourd embellished with white and red feathers that he sent before him by messengers to symbolize authority. This method worked for commanding obedience previously in the western part of Texas on the journey with Cabeza de Vaca.

The messenger sent back to Frier Marcos by Zemmouri told him that he should follow Zemmouri's path after he found people who gave him information of a powerful province. A native from the said province told Frier Marcos that the first city of the province called Cibola was a thirty-day journey from where Mustafa Zemmouri was. The native also confirmed that there were seven great cities in the Province of Cibola under one lord. There the houses were made of lime and stone, some with two and three lofts and the house of the lord was of four lofts. The gates of the principal houses were rimmed with turquoise stones which were of plenty in the region.[xxxv]

Frier Marcos delayed his following of Mustafa Zemmouri waiting for the messengers he had sent to the sea. He expected Zemmouri to wait for him because he had been ordered by the Viceroy to render implicit obedience to the leader of the expedition. However, Zemmouri continued his journey to Cibola without waiting for Frier Marcos. Within a day's journey, Zemmouri sent his messengers with his gourd to the city to notify the chief that he came in peace. When the messengers returned to Zemmouri, they told him that the chief became angry when he received the gourd and threw it to the ground. Mustafa in his ambitious disposition was not discouraged by the news and told the company to fear nothing and he still wished to go there. Zemmouri and his party of three hundred men and many women reached Cibola at sunset. He was not permitted to enter into Cibola but given a house of good lodging outside of the province.

Frier Marcos left on Easter Tuesday on the same route Zemmouri took. Messengers not only told him the first city of Cibola was thirty days out, but there were three other kingdoms beside the seven cities named Marata, Achus, and Tontonteac. Frier Marcos reached a village where he was well received and given knowledge about Cibola. Mustafa Zemmouri left a large cross leaving news of the increasingly good country. The people of the village told Frier Marcos that some of their people left with Zemmouri on a four or five-day journey. Before Frier Marcos made it to the desert, he met people in a village who had turquoise hanging from their nostril and ears. They told him of a cloth made from a little beast that Zemmouri carried with him was in great amounts at Totonteac.

At Cibola, Zemmouri was stripped of all he had–they withheld water and food from his company. Early the next morning, Mustafa Zemmouri left the house accompanied by some of the chiefs. Many townspeople appeared, and Zemmouri fled with his crew. They were pursued and slain.

Two of the natives with Zemmouri who escaped brought the news to Frier Marcos. They believed that Zemmouri had been killed by arrows but did not witness his death. Still the manner of his execution is not certain, and it is said that his body was divided into pieces and distributed among the chiefs. Another version of Zemmouri's fate is that he escaped with the chiefs in his company and hid in the villages to conceal himself from his pursuers and the Spaniards. There is a story amongst the Zuni about the Black Mexican. According to the book The Spanish Settlements, Mr. Frank H. Kushing places the event at Kiakima in the Zuni Plain.

"It is to be believed that a long time ago, when roofs lay over the walls of Kyakime, when smoke hung over the housetops, and the ladder-rounds were still unbroken in Kyakime, then the Black Mexicans came from their abodes in Everlasting Summerland. One day, expectantly, out of Hemlock Canon they came, and descended to Kyakime. But when they said they would enter the covered way, it seems that our ancients looked not gently at them; for with these Black Mexicans came many Indians of Sonoli, as they call it now, who carried war feathers and long bows and cane arrows like the Apaches, who were enemies of our ancients.

Therefore, our ancients, being always bad-tempered, and quick to anger, made fools of themselves after their fashion, rushed into their town and out of their town, shouting, skipping, and shooting with sling-stones and arrows and tossing their warclubs. Then the Indians of Sonoli set up a great howl, and thus they and our ancients did much ill to one another. Then and thus was killed by our ancients, right where the stone stands down by the arroyo of Kyakime, one of the Black Mexicans, a large man, with chili lips, and some of the Indians they killed, catching others. Then the rest ran away, chased by our grandfathers, and went back toward their country in the Land of Everlasting Summer.

But after they steadied themselves and stopped talking, our ancients felt sorry, for they thought, 'Now we have made bad business for after a while these black people of Sonoli Indians, being angered, will come again.' So they felt always in danger from fear, and went about watching the bushes. By and by they did come back, these Black Mexicans, and with them many men of Sonoli. They wore coats of iron, and war bonnets of metal, and carried for weapons short canes that spit fire and made thunder, so said our ancients, but they were guns, you know. They frightened our bad-tempered fathers so badly that their hands hung down by their side like the hands of women. And these black, curl-bearded people drove our ancients about like slave-creatures.

One of these coats of iron has hung a long time in Isleta and there people say you may see it. After that the Black Mexicans were peaceful, they say; but they went away and sometimes came back, it seems, and never finished making anger with our ancients it seems."[xxxvi]

Moorish Castle at Mustafa Zemmouri's birth city of Azemmour, Morocco

Left: *Earthen adobe in Acoma Pueblo, New Mexico* ***Right:*** *Earthen adobes of Morocco*

The Tristan de Luna Expedition

After nearly a hundred years of exploration, war and torture amongst the indigenous people, the Spanish conquistadors still had not established a permanent settlement in North America. From Ponce de León's falling to the Calusa's arrow, to the Ill-Fated Journey of Narváez and Cabeza de Vaca to the ruthless expedition of De Soto, the inhabitants and the land itself proved that they would not be conquered. After the news of the great anchorage of Pensacola Bay, another Spaniard named Tristan de Luna y Arellano was the next explorer to attempt to establish a post at the bay. Pensacola Bay would be a good position for sea communications out to the Gulf of Mexico and Cuba potentially providing trade advantages. In 1559, Tristan de Luna would be the first successor of De Soto to be sent to the Florida coast to establish a post for commercial exchange with access to the interior of North America.

Pedro Hernandez Canillas and Roderigo Ranjel wrote a letter in 1557 to King Philip suggesting that a settlement should be established in Pensacola to provide shelter for ships in distress and a base for the settlement of Florida. Likewise, the inspector Dr. Pedro de Santander wrote a second letter to the king of Spain suggesting that the first settlement should be made in that "most fertile" province of Pensacola, where a new town called Filipina would soon become a center for trade with and conversion of the natives. Dr. Pedro de Santander wrote to King Philip in detail a scheme for colonizing Florida. In his letter, he presents religious justification for the theft of the land and the forced conversion of its people stating:

"This [Florida] is the land promised by the Eternal Father to the faithful, since we are commanded by God in the Holy Scriptures to take it from them, being idolaters, and by reason of their idolatry and sin, to put them all to the knife, leaving no living thing, save maidens and children, their cities robbed and sacked, their walls and houses levelled to the earth."[xxxvii]

In 1558, Guido de Lavazeres explored the northern Gulf Coast. On the journey from Veracruz to Choctawhatchee Bay, Lavares inspected Mobile Bay and named it Bahia Filipina after King Philip. He reported Mobile Bay to be the largest and most excellent bay he had seen, having a large population of natives who grew corn, beans, and pumpkins along the shore. He said that the bay was abundant with game, fish, fruits, and nuts. Lavares was unable to reach the bay of Pensacola due to bad weather. Upon his return to Mexico, he recommended Mobile Bay for settlement. Later in the same year, Juan de Renteria commanded a ship that reached Mobile Bay and then explored Pensacola Bay which they called Polonza. Juan de Renteria's expedition may have factored into the Viceroy Velesco's decision to send Tristan de Luna to land at Pensacola Bay rather than Mobile Bay.

Tristan de Luna, son of the governor of Yucatan, was appointed as captain general of the fleet and governor of Florida. In the summer of 1559, Tristan de Luna embarked on the expedition to Pensacola Bay as a culmination of the previous attempts to colonize the Gulf

Coast. Tristan de Luna left Veracruz with 1,500 people including six Dominican monks, 900 civilians, 500 soldiers and 100 "Aztec warriors".[xxxviii] After a month's travel, the De Luna fleet landed within 27 miles of the Bay of Miruelo on July 17, 1559. One hundred horses aboard the ship had died. Tristan continued westward to locate Pensacola Bay but passed it and moored at Mobile Bay. With the objective still being to establish a settlement at Pensacola Bay, he headed back east and arrived at the Bay of Pensacola on August 14, 1559. De Luna named the new settlement the Bay of Santa Maria Filipina after the Virgin Mary and King Phillip.

The horsemen who were left ashore at Mobile Bay made it back to Pensacola by land. Unlike the landing on Pensacola Bay by Ponce de León and Narváez, Tristan de Luna faced no resistance from natives for they had migrated into the interior after encroachment from past strangers. After landing, De Luna sent exploring parties into different directions to explore the interior and the river that emptied into the Bay of Pensacola. One of these parties probably followed the Escambia River to the north. The Escambia is the largest tributary river of Pensacola Bay. Some of the provisions were unloaded, but the majority of the year's supplies were left aboard the vessel.

On September 19, 1559, a hurricane struck from the north and devastated De Luna's camp, destroying their ships and the supplies aboard. The storm lasted twenty-four hours with increasingly violent winds and resulted in a great loss of life. Two relief ships were sent by the Viceroy in November to sustain the people throughout the winter. De Luna became delirious with fever during the winter and all the food left at Pensacola Bay had been consumed. De Luna with one thousand colonists traveled up the Alabama River to reach the town of Nanipacna where they stayed for months. Nanipacna proved to be insufficient and in the spring of 1560 the colony again wandered for food. The soldiers ate acorns, and the women and children went in search for leaves and twigs. The remaining men of the army opposed De Luna's order to march to the large town of Coca.

When the bad news arrived in Mexico and the Viceroy learned about the outcome of the Spanish settlement he was dissatisfied with De Luna's conduct. He appointed Angel de Villafane to take the place of Tristan de Luna. Four months before he was commissioned by the Viceroy, he had sailed from Veracruz with orders to occupy Santa Elena in South Carolina. When Villafane arrived at Pensacola Bay, he offered any who decided to come on an expedition to Cuba and Santa Elena. Tristan de Luna declined and returned to Mexico where he died in 1571. Villafane would continue and arrive at Santa Elena, now Parris Island, South Carolina on May 27, 1561.

I Love Little Al-Andalus (Pensacola)

From my times shopping in the markets of Cordova

To enjoying the art galleries and festivals of Seville

I love Little Al-Andalus, home is where my heart is

For I was born in the Sacred Heart of Cordova

Seafood and the Sun hold a sacred place in my heart

There in Granada live the descendants of an industrious people

They built independent communities in a time of hate

Just as the Moors built Al-Andalus during the Crusades

The Moors there arranged festivals every day, always in love

A city of champions, Olympic medalists, and legends of sport

A city of five flags, the rich culture of Little Al-Andalus I love

Under all five flags, our ancestors fought for this land

Substantial finances cultivated in the soil by their hands

I love Little Al-Andalus, your oldest neighborhood is Seville

Where Julee Panton's cottage still stands escaped captives once lived

Where Chief Abraham escaped the grasps of Andrew Jackson's army

Later he became a great Seminole leader as Warrior of the Suwanee

I love Little Al-Andalus, may the districts of Cordova, Seville, and Granada prosper

Like Spain's kingdoms of Cordova, Seville, and Granada under the Medieval Moors

I love Little Al-Andalus, from your land, let booming industries be revived

As Booker T. visited and saw it as a place where people of color could thrive

I love Little Al-Andalus, may vibrant cultural venues be reborn

When James Brown visited the Savoy and sang 'Get on Up!'

How Tina Turner came to Devilliers and sang 'Rolling on the River'

Where B.B. King came with Lucille, may the thrill never go away

May the Moors own real estate and their companies flourish

To them let favor, currency and commerce abundantly flow

As Estévanico the Moor came and later found the Seven Cities of Gold

Let our intellectual spheres expand and economic opportunities grow

May the Moorish estates be as vast as the seven seas with acquisitions of great land

As the De Luna estate in Borobia yielded three million gold, Moorish coins yearly

Oh Little Al-Andalus, let the worthy Moors acquire one hundred times more dearly

Cordova, Granada, and Seville; the Moorish essence within the place I find love

Little Al-Andalus give to us wealth, vitality, peace, and prosperity forever more

End of the Alley at Seville Quarter

References

[i] Lane-Poole, Stanley, "The Legacy of the Moors in Spain" Introduction by John Jackson. (Robert Jastrow, pp.118-119)
[ii] Historic American Building Survey: Julee Cottage
[iii] Wills, Clifford W., "Hannibal", p. 36 New York, NY: Infobase Publishing, 2008
[iv] Crow, John A., "Spain: The Root and the Flower", p. 42 New York: Harper & Row, Publishers, 1963, 1975
[v] Chejne, Anwar G., "Muslim Spain: Its History and Culture", p. 8, Minnesota: The University of Minnesota Press, 1974.
[vi] Crow, John A., "Spain: The Root and the Flower", p. 42 New York: Harper & Row, Publishers, 1963, 1975
[vii] Crow, John A., "Spain: The Root and the Flower", p. 42 New York: Harper & Row, Publishers, 1963, 1975
[viii] ibid
[ix] Crow, John A., "Spain: The Root and the Flower", New York: Harper & Row, Publishers, 1963, 1975
[x] Fayyusi, Salma Khadra, "The Legacy of Muslim Spain", p. 759, Leiden ; Boston : Brill, [1994]
[xi] ibid
[xii] Chejne, Anwar G., "Muslim Spain: Its History and Culture", p. 163, Minnesota: The University of Minnesota Press, 1974.

[xiii] Fayyusi, Salma Khadra, "The Legacy of Muslim Spain", p. 759, Leiden ; Boston : Brill, [1994]
[xiv] Abdul-Hassan Ali Ibn Al-Hussain (871-957 C.E.), "Muruj adh-dhahab wa maadin aljawhar (The meadows of gold and quarries of jewels)".
[xv] Al-Sharif Al-Idrisi (1099-1166 CE) "Nuzhat al-mushtaq fi ikhtiraq al-afaq (Excursion of the longing one in crossing horizons)".

[xvi] Ad-dine Abu-l-Abbas Amad Ben Fadhl Al-Umari (1300-1384 CE), "Massaalik al-abasaar fi mamaalik al-amsaar (The pathways of sights in the provinces of kingdoms)."

[xvii] Bovill, E W., "The Golden Trade of the Moors", New York: Oxford Univ Pr, 1958.
[xviii] Hamidullah, "L'Africa decouvre l' Amerique avant Christophe Colomb", pp. 173-83, Abbas Hamdani, An Islamic Background to the Voyages of Discovery, "Legacy of Islamic Spain", pp. 276-277. Leiden; Boston: Brill, 1994.
[xix] Fell, Barry, "Saga America", p. 41, New York: Times Books, 1980.
[xx] Speck, Frank G., "The Nanticoke Community of Delaware", p. 4, New York: The Museum of the American Indian, 1915.
[xxi] ibid, p. 3
[xxii] Colón, Fernando, 1488-1539, "The Life of the Admiral Christopher Columbus / by his son, Ferdinand", p. 14, New Brunswick, N.J.: Rutgers University Press, 1959.

[xxiii] Gordon, Cyrus, "Before Columbus", New York 1971.

[xxiv] Huyghe, Patrick, "Columbus was Last", New York 1992.
[xxv] Weiner, Leo, "Africa and the Discovery of America", New York: A&B Publishers republished 1992, first published 1922.
[xxvi] Lowery, Woodbury, "The Spanish Settlements within the Present Limits of the United States", p. 149, New York; Russell & Russell Inc., 1959.
[xxvii] ibid, p. 157
[xxviii] ibid
[xxix] Bolton, H.E., "Corando, Knight of Pueblos and Plains", p. 16, New Mexico: The University of New Mexico Press, 1949.
[xxx] Clissold, Stephen, "Seven Cities of Cibola", p. 80, New York, C. N. Potter [1962, c1961]
[xxxi] Lowery, Woodbury, "The Spanish Settlements within the Present Limits of the United States", p. 215, New York; Russell & Russell Inc., 1959.

[xxxii] ibid, p. 218
[xxxiii] Lowery, Woodbury, "The Spanish Settlements within the Present Limits of the United States", p. 208, New York; Russell & Russell Inc., 1959.
[xxxiv] Roberts, Calvin, A., Roberts, Susan, A. "New Mexico", pp. 24-26, Albuquerque: University of New Mexico Press, 1988.
[xxxv] Lowery, Woodbury, "The Spanish Settlements within the Present Limits of the United States", pp. 208-209, New York; Russell & Russell Inc., 1959.
[xxxvi] Lowery, Woodbury, "The Spanish Settlements within the Present Limits of the United States", pp. 281-282, New York; Russell & Russell Inc., 1959.
[xxxvii] Lowery, Woodbury, "The Spanish Settlements within the Present Limits of the United States", p. 355, New York; Russell & Russell Inc., 1959.
[xxxviii] The Spanish Presence in NW Florida 1513-1705, University of West Florida

*Front Cover Image Credit: T.T. Wentworth Museum, courtesy of UWF Historic Trust

Printed in the USA
CPSIA information can be obtained
at www.ICGtesting.com
LVHW071154281023
762436LV00001B/7